MILK DAIRIES

70 years ago and now

This book is all about how the dairy business has changed since my grandparents milked by hand. Then the use of milk cans to ship the milk. Then the use of bulk tanks and the tanker trucks to the super dairy where the cows ride a carousel and milk is never touched by human hands from the cow to the factory.

By

Martha Philbeck

goldenpaws@embarqmail.com

Promoted by Gunas Publishing

gunaspublishing@yahoo.com

Martha

I wish to specially thank the following people for letting me visit their dairy operations and take pictures.

The Whetstone Family
The Dawes Family
Fair Oaks Dairy

Beginning our Journey

Our journey will begin with memories from my childhood. Life on a farm was a lot different than it is now. There was no mechanization as we see today. We did not have electricity to begin with. Milking the cows was a big job and a great responsibility. My grandparents milked by hand. The milk was then carried into the house in buckets. A cream separator sat in the kitchen which had to be operated by hand so there was fresh cream every morning which could then be sold or made into butter.

Refrigeration did not come to the farm until 1940. Milk was cooled by putting the cans into cold water. The cows spent the day in the field eating grass. The cats always watched grandma milk and waited patiently for their bowl to be filled.

Hay and straw were stored in the loft in the barn. A hallway led to the ladder that was always there and which you had to climb to get to the hay to throw it down. The hay was then put in the manger.

This was the regular food for the cows. Their long tail was good to swat flies with.

Their ears are very flexible and can move in many directions to warn them of danger. They knew their name and would come when called. When they are full of milk, they wait impatiently to be milked. Too long a wait can make them uncomfortable. They are obviously more comfortable after milking.

Some cows have horns and some are polled. The horns are made of the same material as your fingernails. Most farmers remove their horns for safety reasons. Bull's horns can grow longer than females. The horns grow on each side of the forehead.   A cow has many parts.

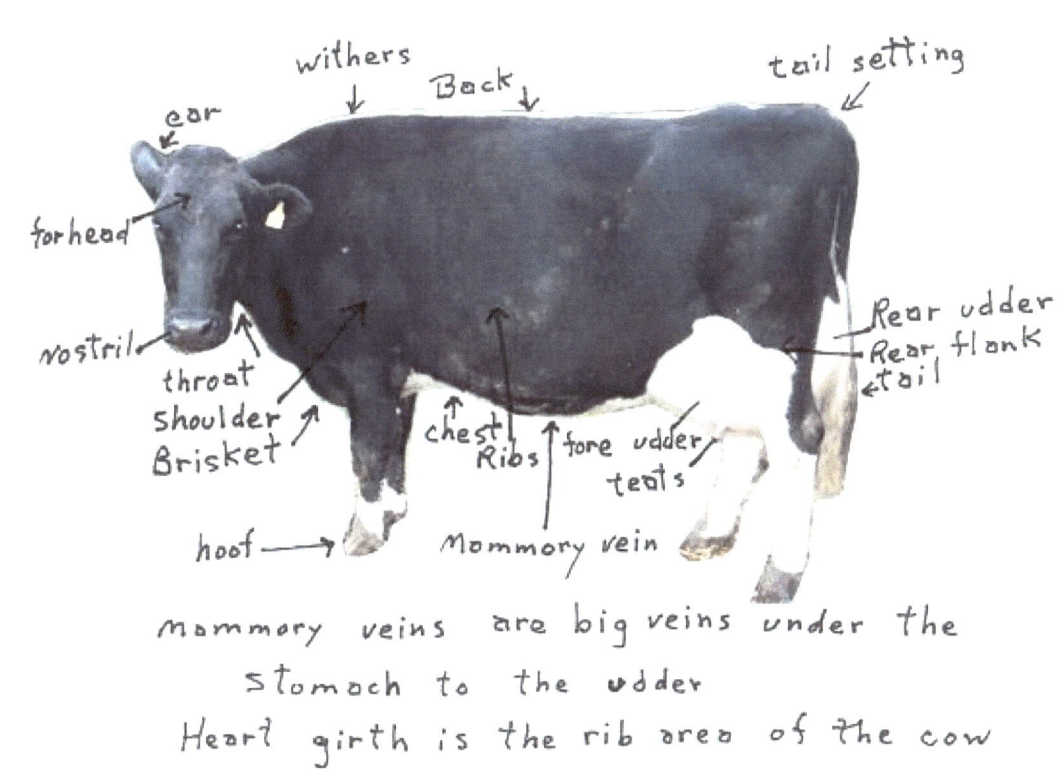

mammory veins are big veins under the
stomach to the udder
Heart girth is the rib area of the cow

The first cow came to America in 1611. Until the 1850's nearly every family had a cow. When it was time to get the cow bred, my grandparents would walk the cow to the neighbors that had a bull. They would milk the cow for 10 months and then have her go dry for 2 months before she was to have another calf. Going dry meant that she would not be milked and therefore would quit producing milk feeling there was no need for it.

In this book we shall be visiting 3 different dairies. We will start out with milking to the super milking stations that are the thing of the future. You will be able to see the difference.

These cows are on pasture in the day time. When it is dry weather they stay pretty and clean. When it rains they do get muddy just like all of us when tramping in the mud.

These cows are waiting to come in and be milked. They know that I am new around there and strange. Cows get to know the people taking care of them and are spooked around strangers. Just like children.

The cows that are not being milked are kept in a different area.

The cows are brought in and fed. Because they can be moving suddenly and this can be dangerous a chain is fastened so they don't move while being milked. The green bowl is for their water.

The cans are washed and put on the rack upside down to drain. They have to be clean and dry for the next time. Everything has to be sterilized. The cans get discolored on the outside from sitting in the cold water to chill the milk. Each farm has its known identifying number and that is marked on the can.

Next the udder is washed and the "milkers" attached. The milkers do the job today that my grandparents did by hand.

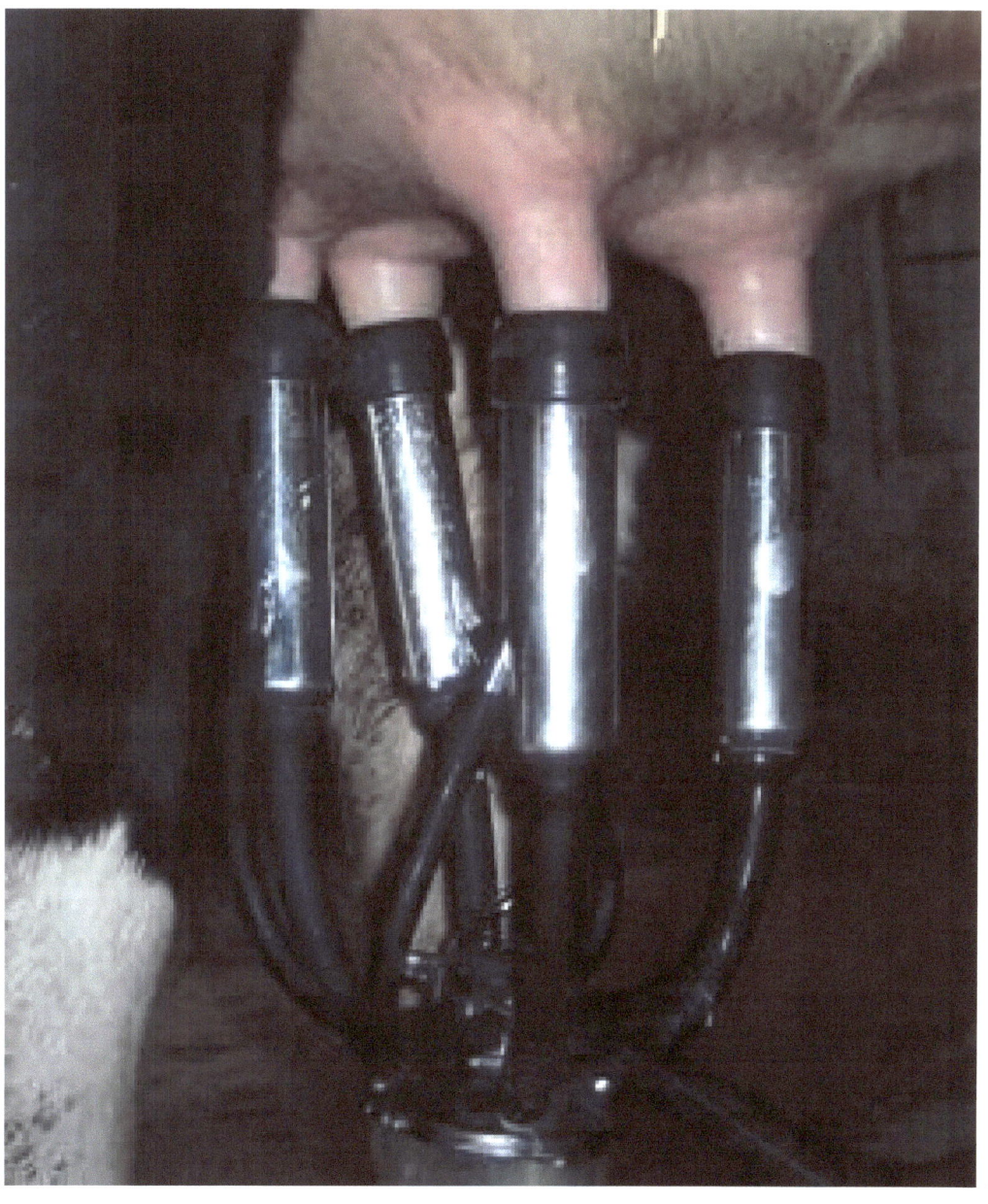

The milk goes into a stainless steel can. Then it is poured into a bucket and weighed. Each cow has a name and a number on her ear. This identifies her. A record has to be kept of how much milk each cow gives.

The "milker" is dipped into a sterilizing solution to make it free from bacteria and infection in between using it on each cow. You can see a bucket in the background with the milker in it waiting for being used on the next cow. The cow's teats are also dipped

into a cleansing solution after milking.

The milk is poured into a funnel sitting on top of a milk can.

The cooling tank is in the background, filled with cold water to place the milk cans in. The milk must be chilled as soon as possible otherwise it will spoil and start curdling.

There is a filter at the bottom of the funnel that would collect anything that is not supposed to be in the milk.

This picture shows the number on the can to identify the farm and the milk in the can. When the can is full the lid will be put on and the can set in the tank of water behind it. This milk is destined to go to a cheese factory.

No farm would be complete without showing a bottle baby.

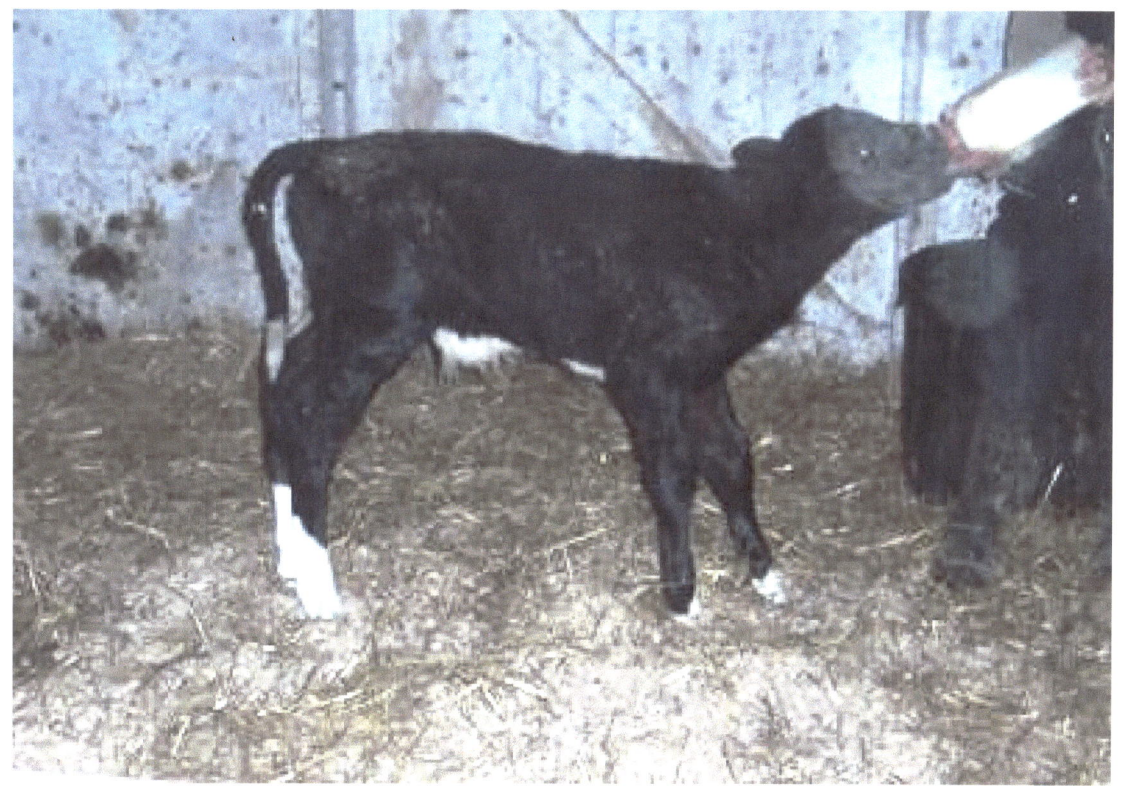

The family farm

Farming is dear to my heart where several generations have worked together. Milking is the hardest since someone has to be available twice a day to make the cows more comfortable. When they fill up with milk, the udder gets very tight. Dairying has improved greatly through the years. Cows today can produce much more milk. One cow can give as much as it once took ten cows.

Cows do not sweat. They can withstand the cold, but do not like the wind. They have cloven hooves which spread and help them to keep their balance in the mud. The pilgrims brought the first cows to America.

The cows enter by a door which you can see behind the last cow. Each side holds 5 cows at a time. They are not fastened in, but are only here until the milkers are done. Holsteins make up 93 percent of all dairy cows. Each one is marked differently. They are usually black and white, with an occasional one with some red; red gene being recessive. It will show up occasionally then you will have a red and white Holstein.

The udders are washed to remove all mud. Here the farmer is holding a hose that is spraying the cow's udders.

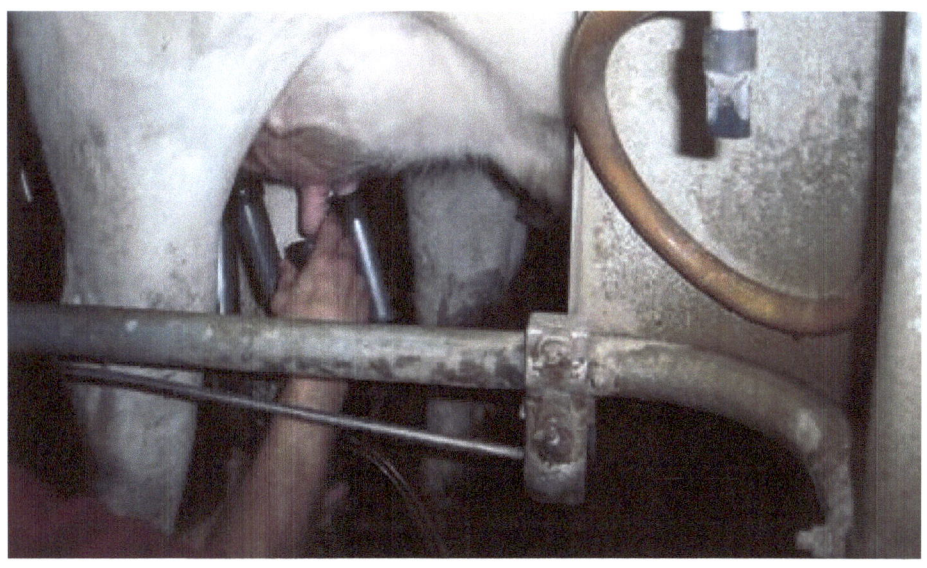

He is attaching the milker which sucks all the milk out of the teat. It applies a gentle air pressure through the brown tube. Works like a calf sucking.

The cows are waiting patiently for the milkers to stop which takes about 5 minutes.

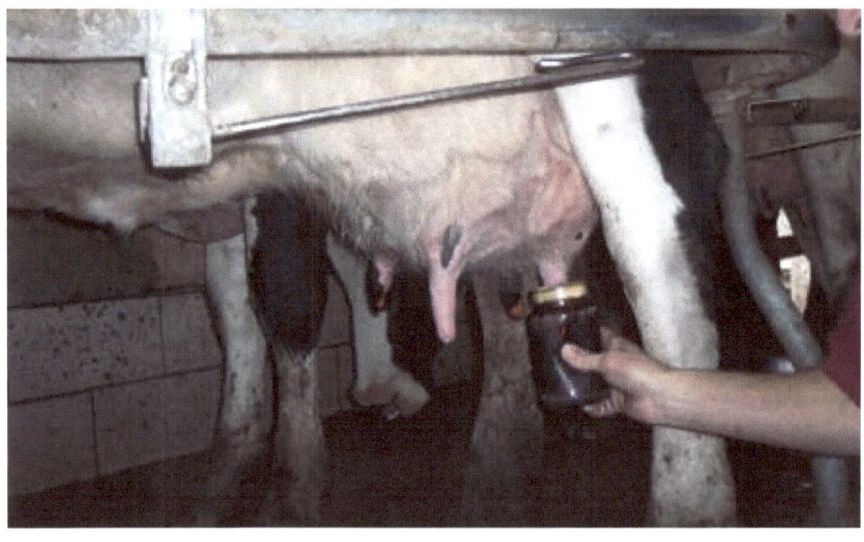

The teats are being dipped into a special solution after milking to seal over the end and prevent infection. This is an antiseptic solution. It takes the teats about an hour to close after milking.

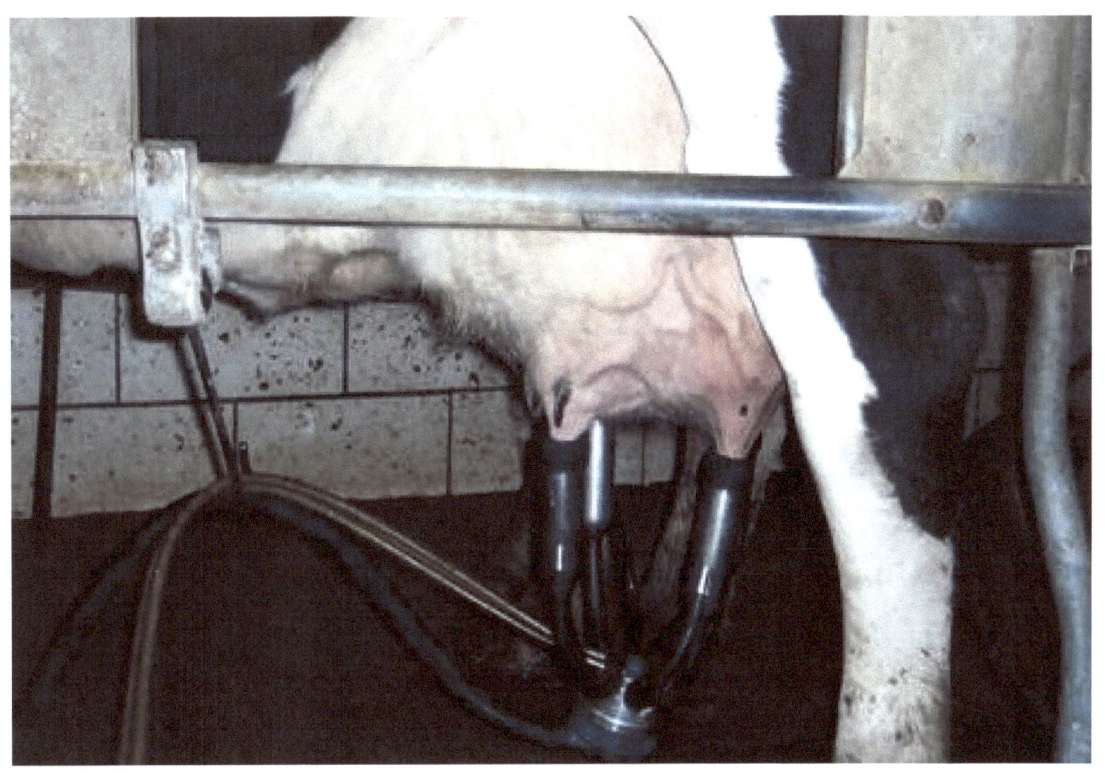

Notice the big veins in this udder. Milk is produced in the udders and comes out of the teats. Tiny cells remove water and nutrients from the blood and convert it into milk; the milk forms into droplets and drips into the udder where it stays until milking time. When the teat is squeezed it will shoot out a stream of milk. There are 4 chambers in the udder and each one has a teat.

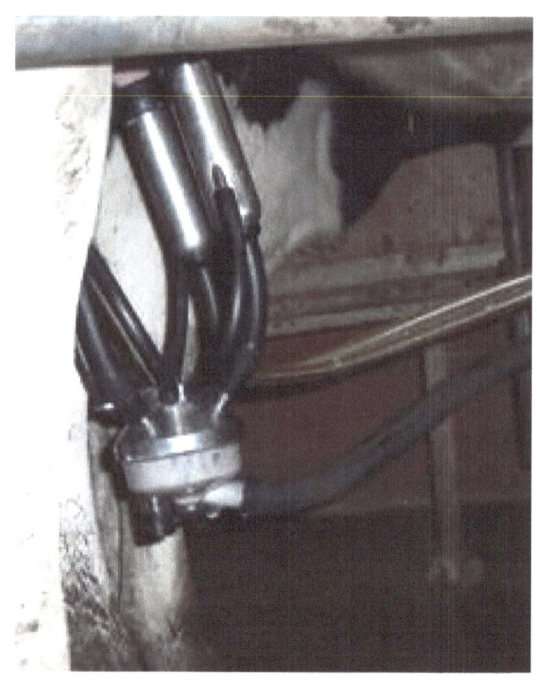

In this picture of the milker, there is a see-through clear portion. You can see the milk in it that is being pulled from the cow. It then goes through the lines to the bulk tank which cools it in minutes.

The calves are taken away several hours after birth and put in pens by them selves. They are given bottles for several weeks and of course they then get in the habit of looking forward to the visit of the person that feeds them.

The first milk that a cow gives after giving birth is called colostrums. It very important for the calf as it is full of substances that give natural immunity to the calf against illness and the new atmosphere it has been born into.

The calves are also given a number so they can be identified with their mother. The number is the tag on their ear. They will always wear that.

The cows are fed hay in the winter when the pasture is not growing. Straw is stored in the loft for bedding.

The big tanker truck comes to pick up the milk. The man from the truck connects a hose to the bulk tank and operates the pump that sucks the milk into his truck. They come every other day to get the milk.

A farm scene from long ago.

Showing Calves at the Fair

By Makenna Dawes

Showing calves at the fair is so much fun. I love it. We start leading our calves in March. When it gets closer to fair time we Wash them

At the fair I like to hang out with my friends and cousins. We have to wash our calves and clip them.

One of the things we do is poop scoop. Every time they poop we have to clean it up with a pitchfork. Then we put it in a wheelbarrow. When the wheelbarrow is full, we have to go and dump it.

Dad makes sure we do it right.

On show day we have to wash the calves. We have to watch them really close so they don't sit down in poop. We braid their tails so when we undo them their tails are fluffy.

I was really scared in my first year. I didn't want my calf to get spooked or anything like that. I get nervous every time I first walk in the ring.

The practice leading has paid off. They are doing a good job.

The judges make us line up. The fair lasts for one week. This coming summer, 2010 will be my third year in 4H. Showing calves is a lot of hard work, but showing them is fun even though you get tired.

Everyone is a winner.

Different Breeds of Dairy Cows

Guernsey

Holstein

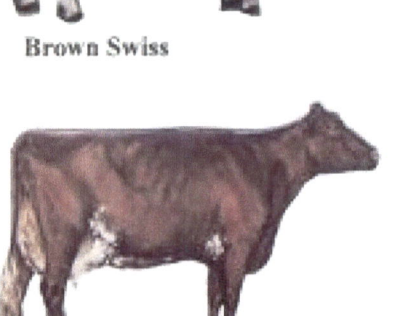
Brown Swiss

Jersey

Milking Shorthorn

Ayrshire

Cows do not have 4 stomachs but they do have 4 digestive compartments; each part is used for a different process. The rumen is a like a large fermentation tank and is found on the left side of the stomach. It can hold up to 50 gallons of fermenting plant matter depending on the size of the cow. This is where the cud comes from.

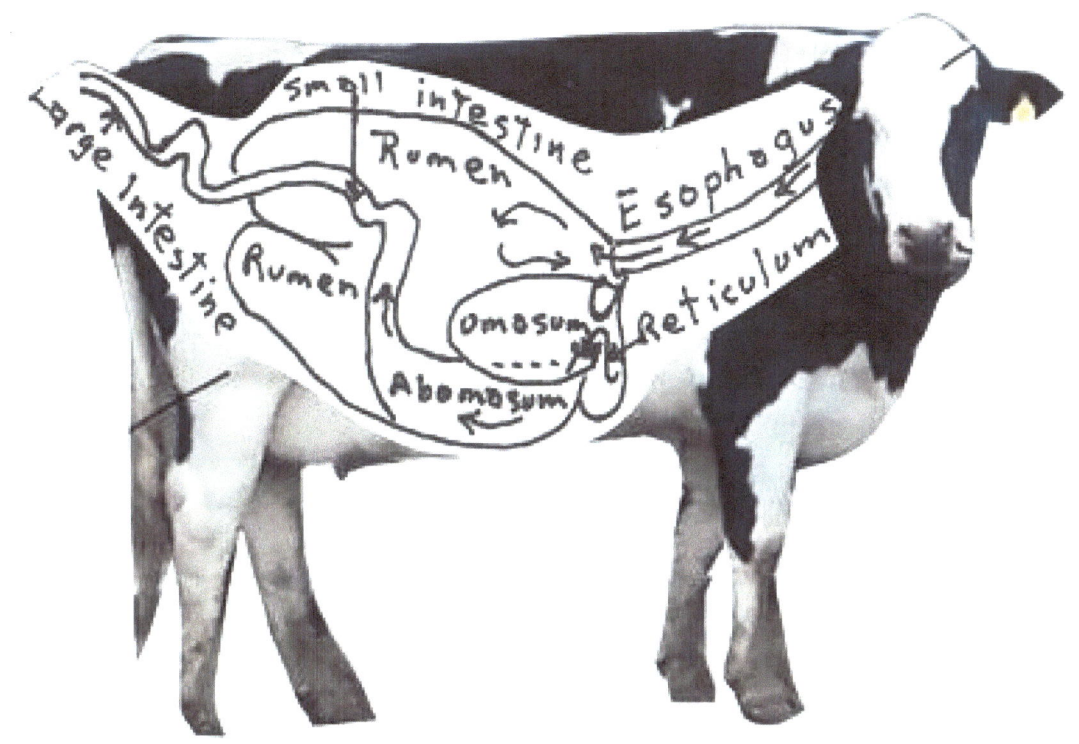

The reticulum is separated from the rumen by a small partition of tissue. About every 30 to 60 seconds the contents of these 2 compartments are mixed together. The breakdown of the food is a slow process. It can take anywhere between 20 to 48 hours. This stomach is called the hardware stomach. If a cow eats something she cannot digest it will stay here and cause no further damage.

The omasum is like a filter. It is made of numerous flaps. It has a large surface area and absorption capacity. Water and minerals are recycled here for a return to rumen as saliva. Over 20 gallons of saliva a day is generated.

The abomasums is a stomach like ours. It uses strong acids and enzymes to break down proteins.

Cow facts

Cows can drink a bathtub full of water and eat 40 pounds of food a day.

Cows are ruminates. These are hoofed animals with an even number of toes.

All ruminates belch up their cud and re-chew their food. This breaks up the plant fibers into smaller pieces and puts more saliva into the food so that it will digest. The cud is a ball of previously eaten grass and other food they have eaten. They swallow their food without chewing it too much at first.

A cow chews her cud for up to 8 hours. When you see a cow just standing around doing nothing but chewing that is what she is doing. Sheep and camels are also ruminants.

Over the years cattle have served man a triple purpose. They provide meat, milk and labor. Later the labor was replaced by draft horses and then machines.

Before milking machines were invented in 1894, cows were all milked by hand. A farmer could milk about 6 cows an hour. With modern milking machines more than 100 cows an hour can be milked.

A young female is called a heifer and she becomes a cow when she has her first calf at an average age of 2. Calves are fed milk until they are 8 to 9 weeks old. They have to have a calf before they will give milk.

A cow has 32 teeth, 8 incisors on the bottom front,   6 molars on the top and bottom of each side; a tough pad of skin instead of teeth on the top front.

A cow's mouth is made for grazing. She tears grass from the field and grinds it between the two mouth parts. Cows are herbivores which mean they do not eat meat, just plants, grass and grain.

Cows could be called the foster mothers of the human race because they produce most of the milk that people drink.

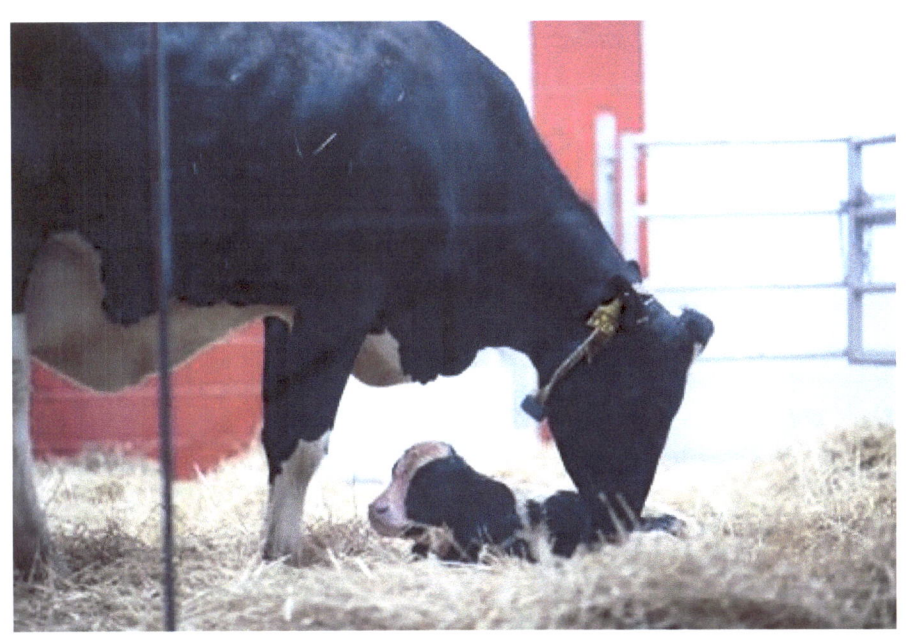

Today we begin our day at the super star of dairies. This is replacing the family dairy business. Nine families went together and started this. They milk 3,000 cows at each of 10 farms three times a day.

We will first go to the birthing barn which is set up for the public to be able to see the miracle of birth. They have 8 to 10 calves born a day.

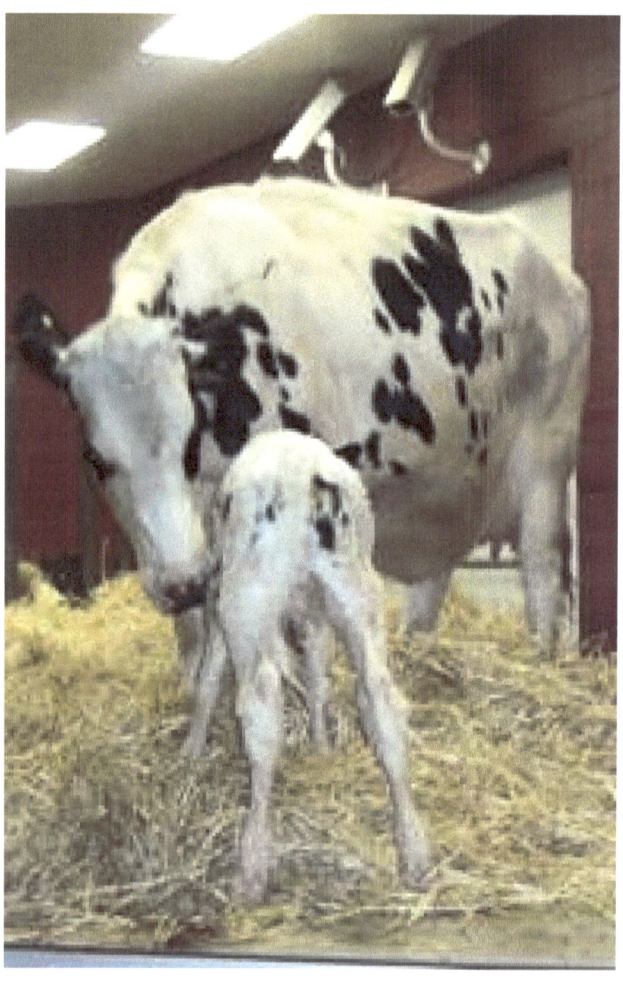

The mothers lick their babies clean and it also helps to stimulate the babies and encourages them to get up on their feet. They are very wobbly at first and have to find their strength. It is a

30

tiresome journey being born. A Holstein calf weighs 80 to 100 pounds and a mature cow weighs 1,300 to 1,500 pounds. Each Holstein has unique markings.

They are given a bottle soon after birth. The first milk produced by a new mother is called colorustrum. It is full of antibiotics and vitamins that will keep it healthy. The mother's milk cannot be used for 3 days by humans. Calves are taken away from their mothers right away.

The average body temperature of a cow is 101.5 F and that is the temperature that milk comes out. It has to be cooled rapidly to prevent bacteria spoiling it.   The bulk tanks cool it to 34 degrees F in minutes. The milk is stored in refrigerated tanks until the tanker truck comes and picks it up.

The milk is later homogenized and pasteurized.

Homogenized means the fat is broken up and mixed into the rest of the milk. The cream comes out of the cow last.

Pasteurization means it is heated to 145 degrees to kill bacteria.

Other animals that are also milked are buffalo, camels, goats, sheep, horses, and reindeer. Dairy cows give 90% of the milk.

As you continue your bus tour of the farm you see the cows in the long building eating. There is always fresh food. The electricity for the farm is made from Manure.

A Jersey calf weighs around 60 pounds at birth. They are one of the smallest cows. They remind me of a baby deer the way that they are marked.

Food is stored outside. It is covered with large tarps and tires are used to hold it down and keep the wind and weather off of it. There are different mixes of foods depending on the needs of the cows.

Picture of the cows walking to the milking parlor. The cows walk to the milking parlor; 500 of them at a time. They patiently wait their turn. This picture shows the cows in the path coming into the milking parlor waiting their turn to come in.

This is a good view of the carousel. The cows walk on the carousel and off without help. Their udders are washed and the milkers attached. It takes 8 ½ minutes to make a round on the carousel. It takes about 5 minutes for the milkers to milk a cow. The cows have a number in each ear to identify them. Records need to be kept on each cow. Their teats are dipped after they finish being milked. Then they get off and walk back to the feeding barn. The milk is taken to the bulk tank by the big white pipes coming down the middle of the picture. Everything is washed and sterilized in the hour between each milking cycle.

This farm also has a gift store and makes cheese and ice cream. Several kinds of the cheese are pictured and you can also watch the cheese making process. It is definitely a trip worth taking.

Making Butter

Milk is made of two parts. Non-fat milk and cream. The cream will rise to the top after it sits. It is much lighter than the milk and is what you use to make butter.

This milk has been sitting for several days. You can notice the difference in color of the top inch of milk. That is the cream. We carefully scooped it out and there was a cup of cream. We will show two different ways of making butter. We let the cream sit all day before churning.

There were different ways of separating the cream. My grandmother had a blue cream separator sitting in her kitchen. The milk was brought in each morning and poured into the separator.

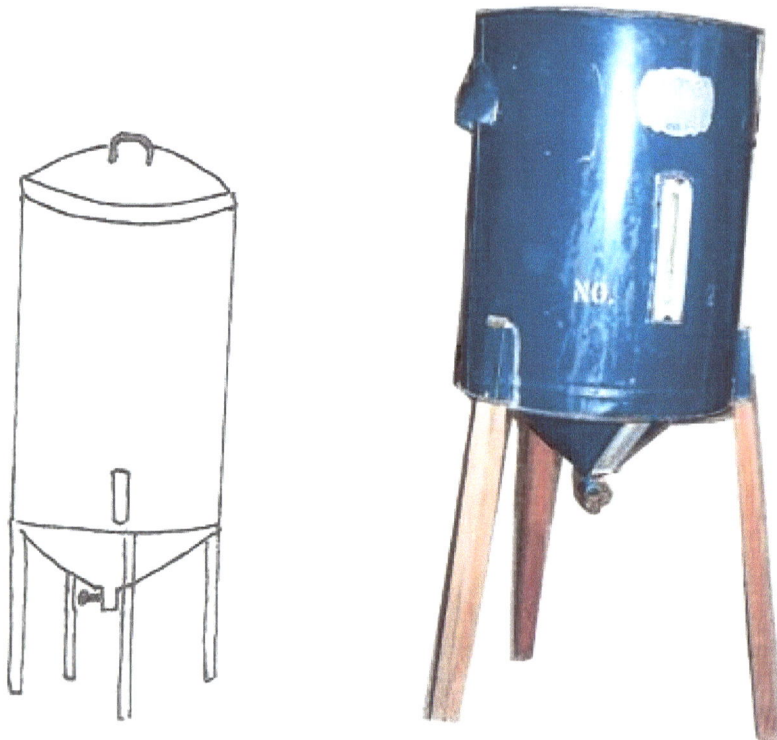

The one on the left I drew from memory. The one on the right a friend owned and it has been passed down thru the generations. It is missing the lid so I included both pictures. The lid was lifted and the milk poured in. It sat on 3 or 4 legs, tall enough to place a bucket under it. There was a window in the front to tell if the cream had risen. Then the spigot was opened and the milk drained off leaving the cream. After the milk was taken out, the cream was drained into another container. A jersey cow is known for giving milk with the most cream.

This is a centrifugal cream separator. This one was originally hand-cranked but later on electric motors were added to it. The milk was poured into the top and went through a series of disk and then the milk came out one side and the cream out the other of the silver spouts. This was the only kind of cream separator that would separate goat's milk.

We used a one quart Dazey Butter Churn. This one is very rare. It has a crank and paddles that stir the cream. There are other styles of churns also. A churn is a thing that is used to make butter at home or on the farm. A regular canning jar can be used, just add your cream and shake it back and forth. It takes awhile. Shake about once a second. Add a marble in it to speed things up. Kids love to do this, but expect it to take between 5 and 30 minutes, depending on how they shake the jar.

There were a lot of hand turned glass churns made, but the Dazey was the best known. The first ones came in 1, 2, 3, and 4 quart sizes. Later the 6 and 8 were added. After 1936 most of the wooden paddles only had two blades. Most families made their own butter. At this time you could not go to the store and buy butter and during World War 1 there was a shortage of butter.

If you want to culture the cream before churning, add a few tablespoons of store bought cultured yogurt, buttermilk, sour cream, clabbered cream or crème fraiche(fresh cream). Let this sit about 12 hours at warm temperature about 75 degrees. It will thicken and ferment before churning. It should taste delicious, slightly sour and have no aftertaste. Don't use it if it is bubbly, or smells yeasty or gassy.

A kitchen scene from long ago

You can use heavy whipping cream or double cream without stabilizers to make butter.

You can also use a food processor or a mixer. The cream will go through the following stages, slushy, frothy, soft whipped cream, firm whipped cream, coarse whipped cream. Then suddenly the fat will all seize together and the sound of the machine will change. A few seconds later a glob of yellowish butter will separate from the milky buttermilk. Drain the buttermilk off. You will get about half as much butter as the amount of cream you started with.

Traditionally they made butter from soured cream. Since we are using electric appliances, we do not need to sour the milk first. If you want a stronger flavor, you can add a tablespoon of store bought cultured buttermilk to each cup of cream used. Let it sit for 12 hours at room temperature before beginning the process.

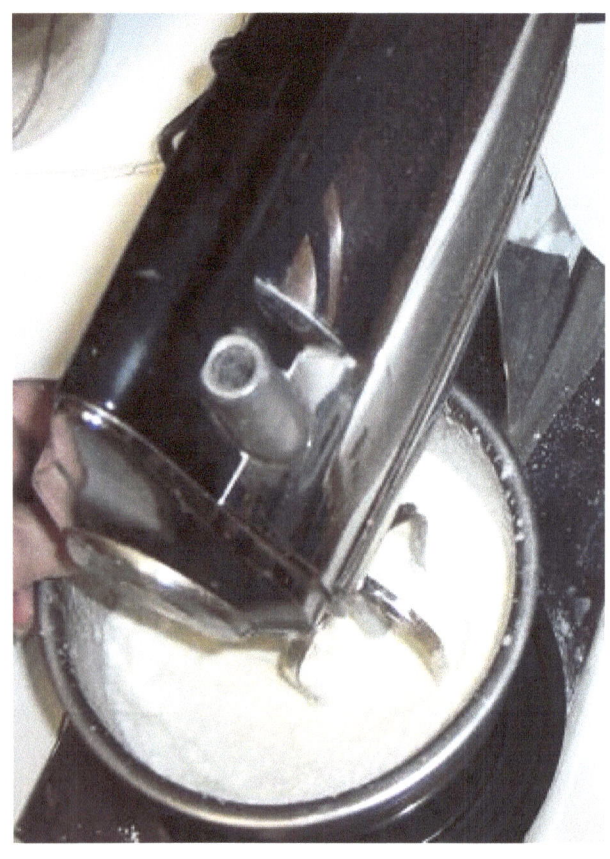

Pour the cream in the bowl of the mixer. Use a big enough bowl that the cream will not slosh out when the mixer is started. Start mixing the cream slow. This cream has reached the whipped cream stage. Increase your speed if it will allow without splattering. In just a little while it will form soft peaks. The next stage will be stiff peaks.

You can tell in this picture how the cream is starting to clump up. It is also a light yellow in color. In just a few seconds it will be butter. This happens quickly. The cream will suddenly seize and be floating on top of the buttermilk. You will need to slow down your speed at this point or you will be throwing buttermilk all over the kitchen.

At this point, drain off the buttermilk. I like to wash the butter.

You can do this by pouring some ice water over it, pushing it around with the spatula and pouring it off. I do this several times. When I was a kid we had a special butter paddle for this job. While the butter is still soft you can add a little salt if you want. Salted butter can be made by whipping ¼ teaspoon table salt to every 4 ounces of butter. You can also make flavored butter by adding herbs and garlic. Use about 1 clove garlic finely minced for every 4 ounces of butter. You can also add about 2 tablespoon of dried herbs for every 4 ounces of butter.

Now that your butter is ready, you can wrap it and slip into the freezer or the refrigerator.

Other Books by Martha Philbeck

THE GOLDEN BOOK OF GOLDEN RETRIEVERS

A compendium of their History, Raising and Training

By Martha Philbeck

A lovely story of a goose that found it difficult to take off but did not give up. By sheer dint of will power and application he did finally manage to fly; of course, with the help of his father!

The Goose that could not Fly

Promoted by:
Gunas Publishing
gunaspublishing@yahoo.com

US$

By Martha Philbeck

Martha's *Farmhouse Cuisine....*

Martha's Special collection of

Countryside Desserts...

Recipes from the 1900s for the kitchens of 2000s

By Martha Philbeck

Martha's *Farmhouse Cuisine....*

Recipes from the 1900s for the kitchens of 2000s

**Collection of Main Dishes,
Farmhouse Wisdom and much more**

By Martha Philbeck

This is the story in easy steps of the life and family and birth and nurturing of the Canadian goose. In easy to understand language you learn about the habits and needs of the Canadian goose and how to raise the young. Full knowledge about the eggs and their incubation is also included.

THE CANADIAN GOOSE
BY MARTHA PHILBECK

Promoted by
Gunas Publishing
gunaspublishing@yahoo.com

Charlie Coyote and other short stories.

Martha Philbeck & others

WILDLIFE ON MARTHA's FARM

The farm is a place of solace where we can connect with nature and vibrant life.

Pictures and texts by Martha Philbeck.